# GET READY TO ENTER

## THE AMAZING WORLD OF DINOSAURS!

The word **'DINOSAUR'** was created by an English scientist called Sir Richard Owen in 1842. The word means **'FEARFULLY GREAT LIZARD'**.

Over **700 different types** of dinosaurs used to roam all over the world, in every continent and every ocean. Some were as small as birds and others as big as buses! Some just ate plants, **these dinos were called herbivores.** Some ate meat including other dinosaurs, **they were called carnivores**. Some ate both plants and meat and **these were called omnivores.**

Dinosaurs lived on earth for about **200 million** years before most died out **(they became extinct)**, 65 milion years ago! We don't know exactly why this extinction happened. It might have been due to an asteroid hitting the planet that changed Earth's weather, or because of rising sea levels or erupting volcanoes.

After the big dinosaur extinction, the avian (bird-like) dinosaurs **evolved into the birds** we see around us today, so you can actually see dinosaur relatives in your garden right now!

Scientists who study dinosaurs are called **Paleontologists**. These experts study **fossils** of dinosaur bones and footprints to understand how and when the dinos lived and what they looked like. **New fossils are found all the time**, there might be some to discover near where you live!

Until then, enjoy coloring and learning about all the dinosaurs in this book!

# BRONTOSAURUS

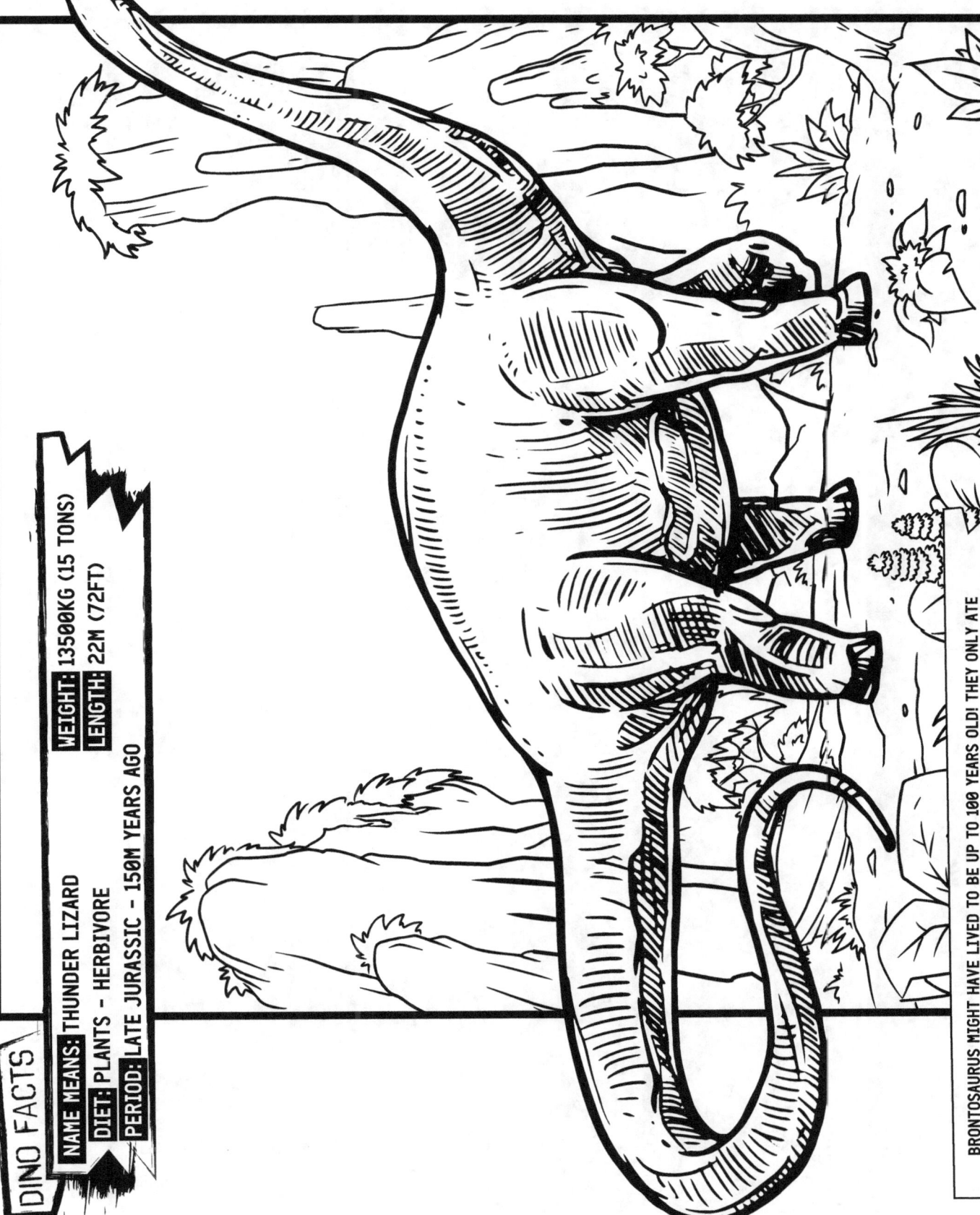

## DINO FACTS

**NAME MEANS:** THUNDER LIZARD
**DIET:** PLANTS - HERBIVORE
**PERIOD:** LATE JURASSIC - 150M YEARS AGO
**WEIGHT:** 13500KG (15 TONS)
**LENGTH:** 22M (72FT)

BRONTOSAURUS MIGHT HAVE LIVED TO BE UP TO 100 YEARS OLD! THEY ONLY ATE PLANTS BUT MIGHT HAVE EATEN STONES TOO TO HELP THEM DIGEST THE PLANTS THEY DIDN'T CHEW.

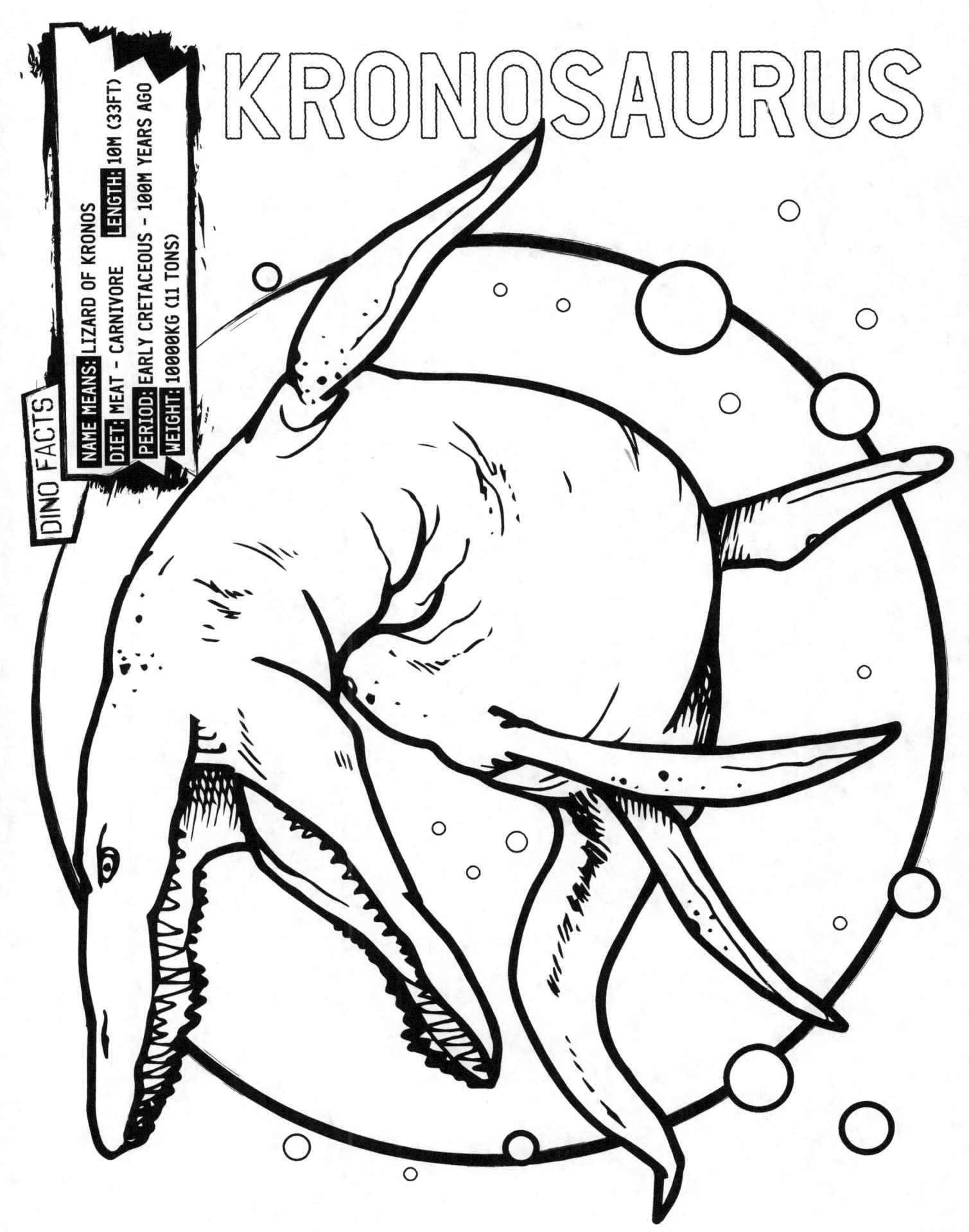

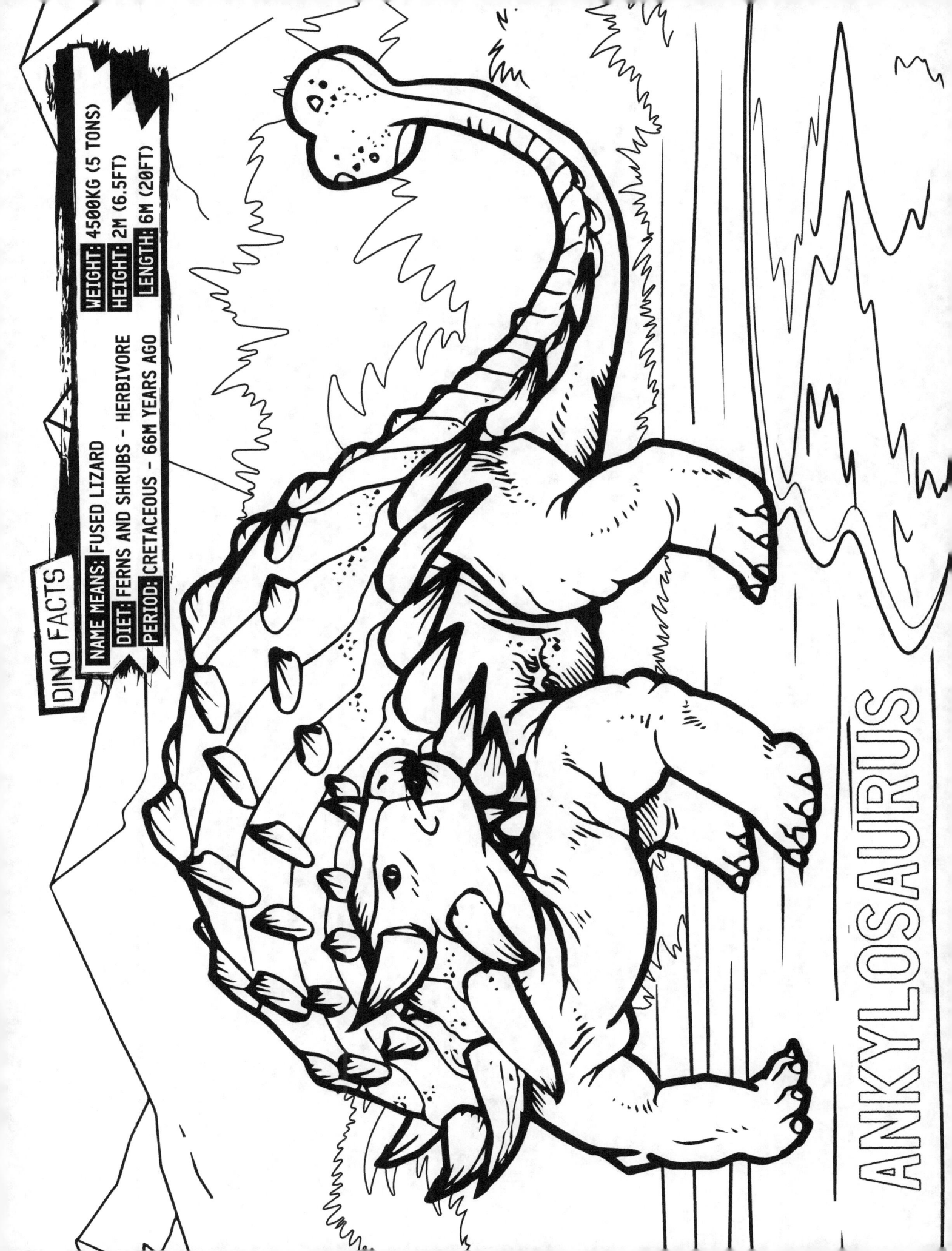

VELOCIRAPTORS HUNTING

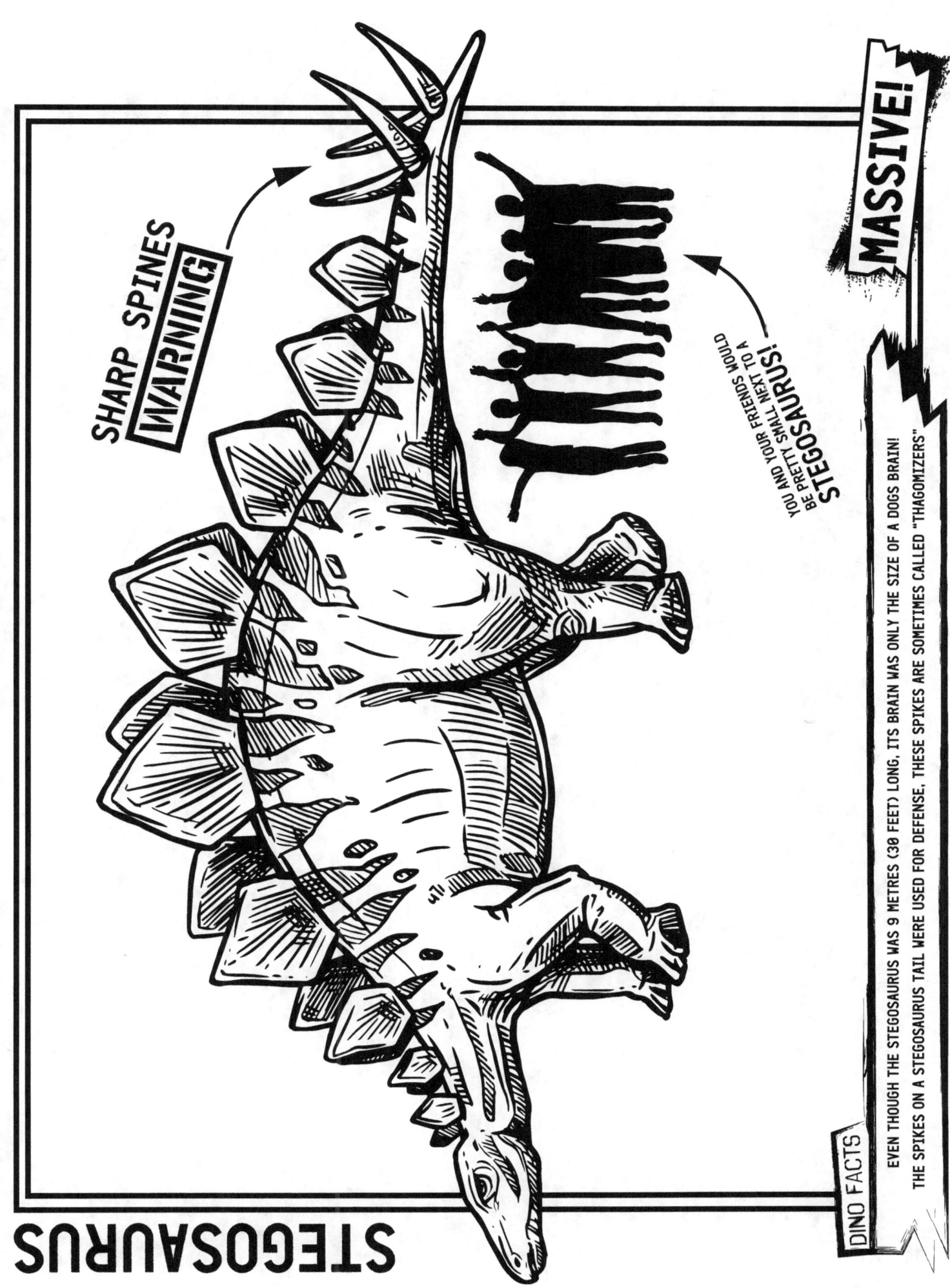

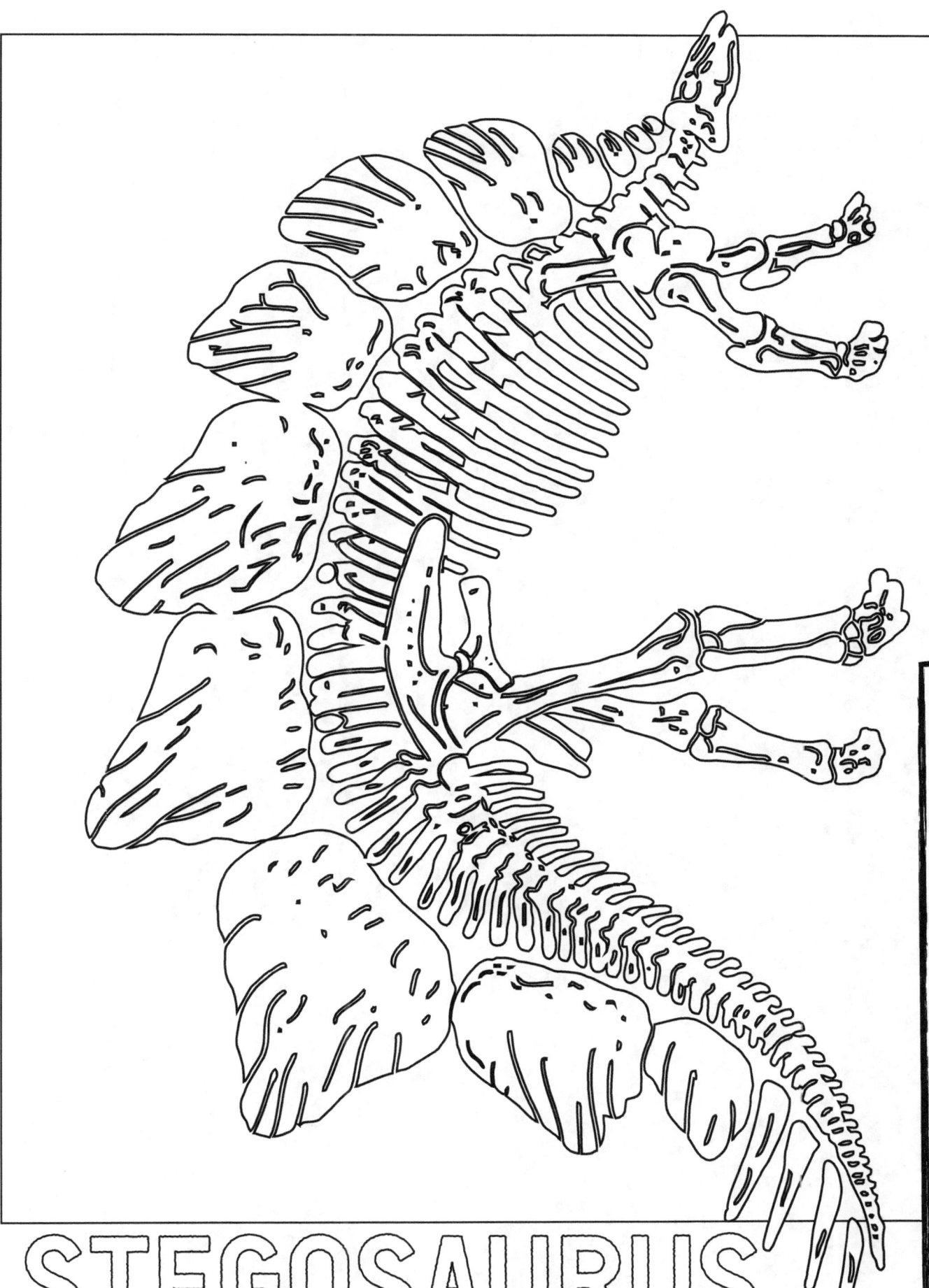

# STEGOSAURUS

THE FIRST STEGOSAURUS FOSSIL WAS FOUND IN COLORADO, USA. BECAUSE OF THIS IT WAS NAMED THE STATE DINOSAUR OF COLORADO IN 1982.

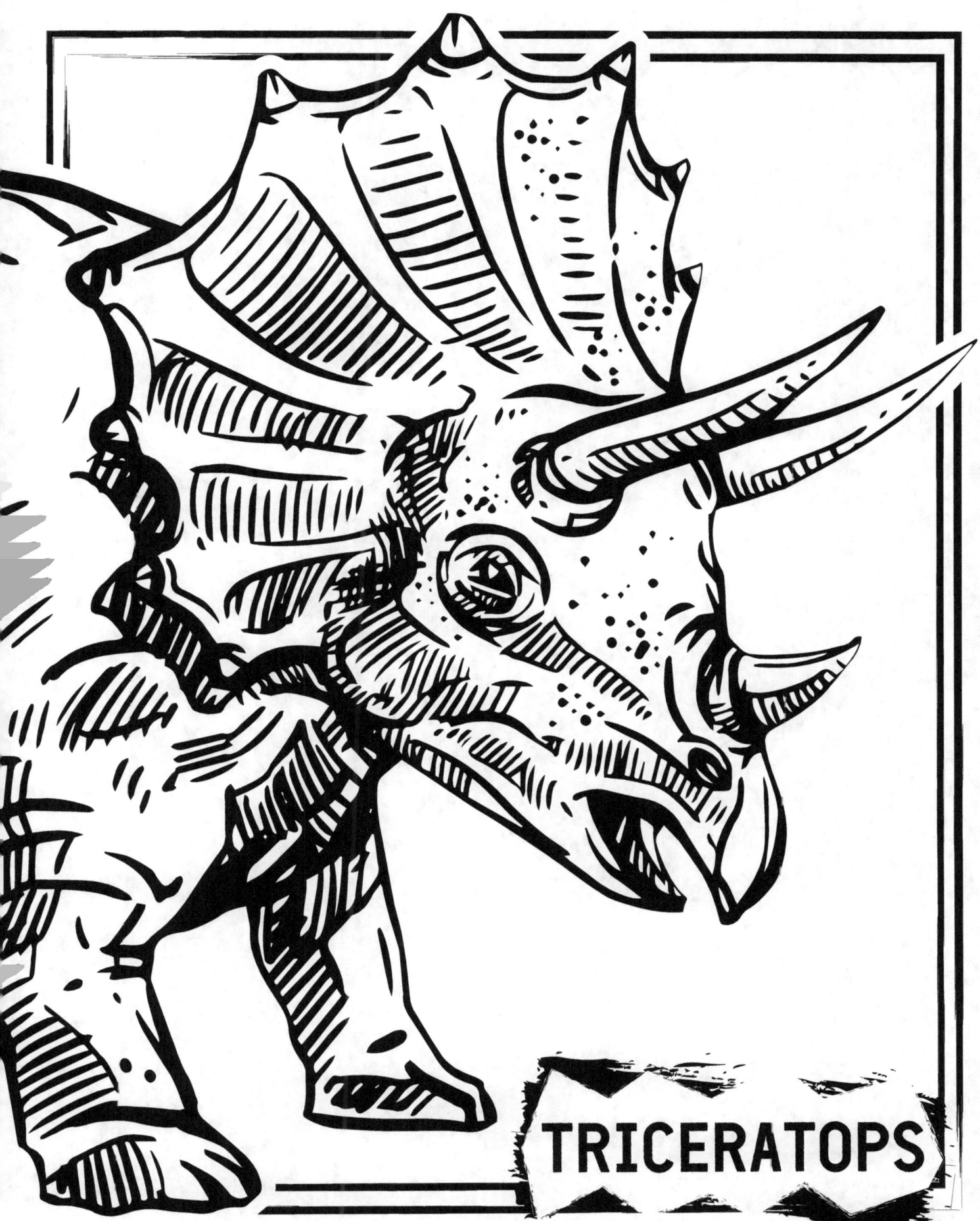

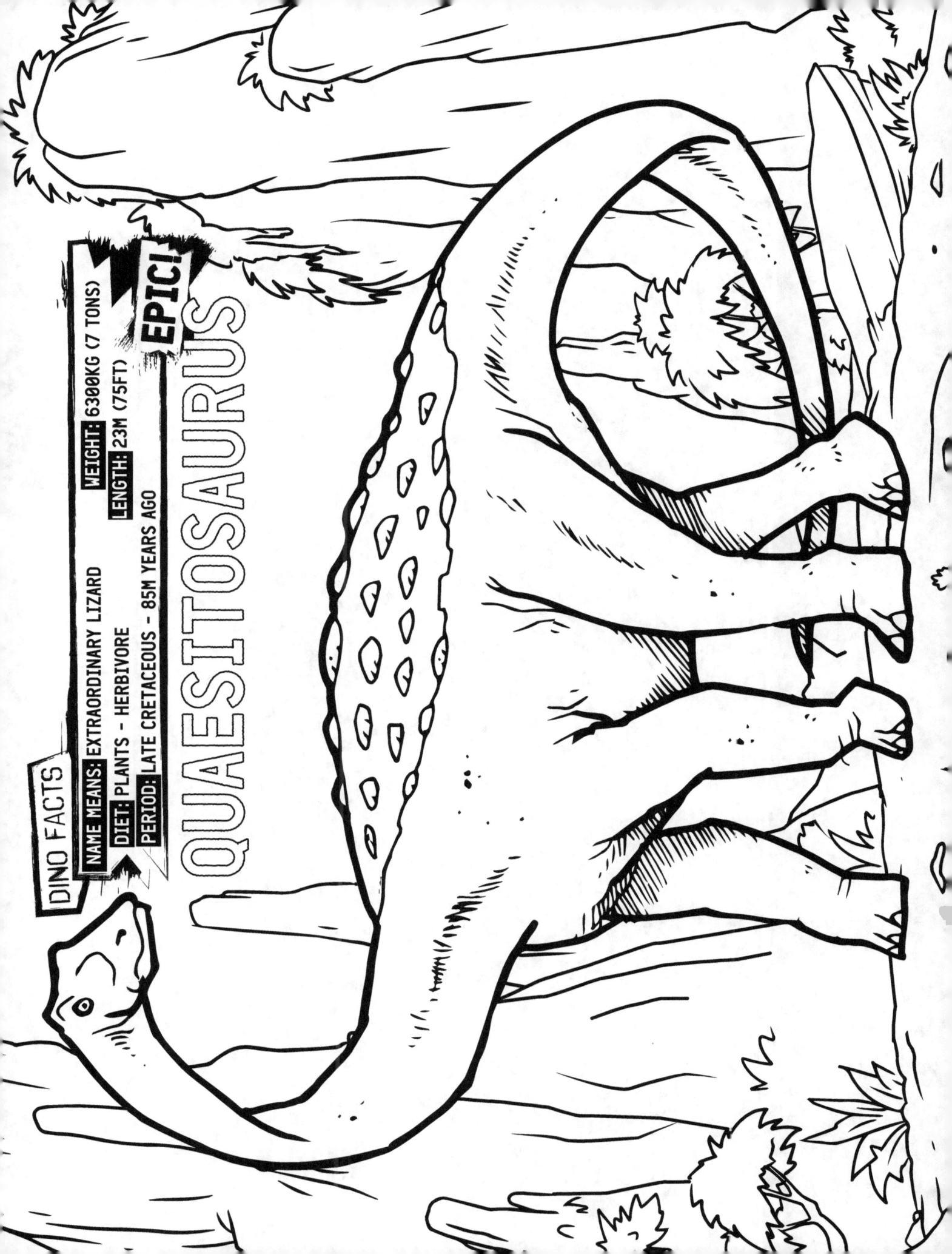

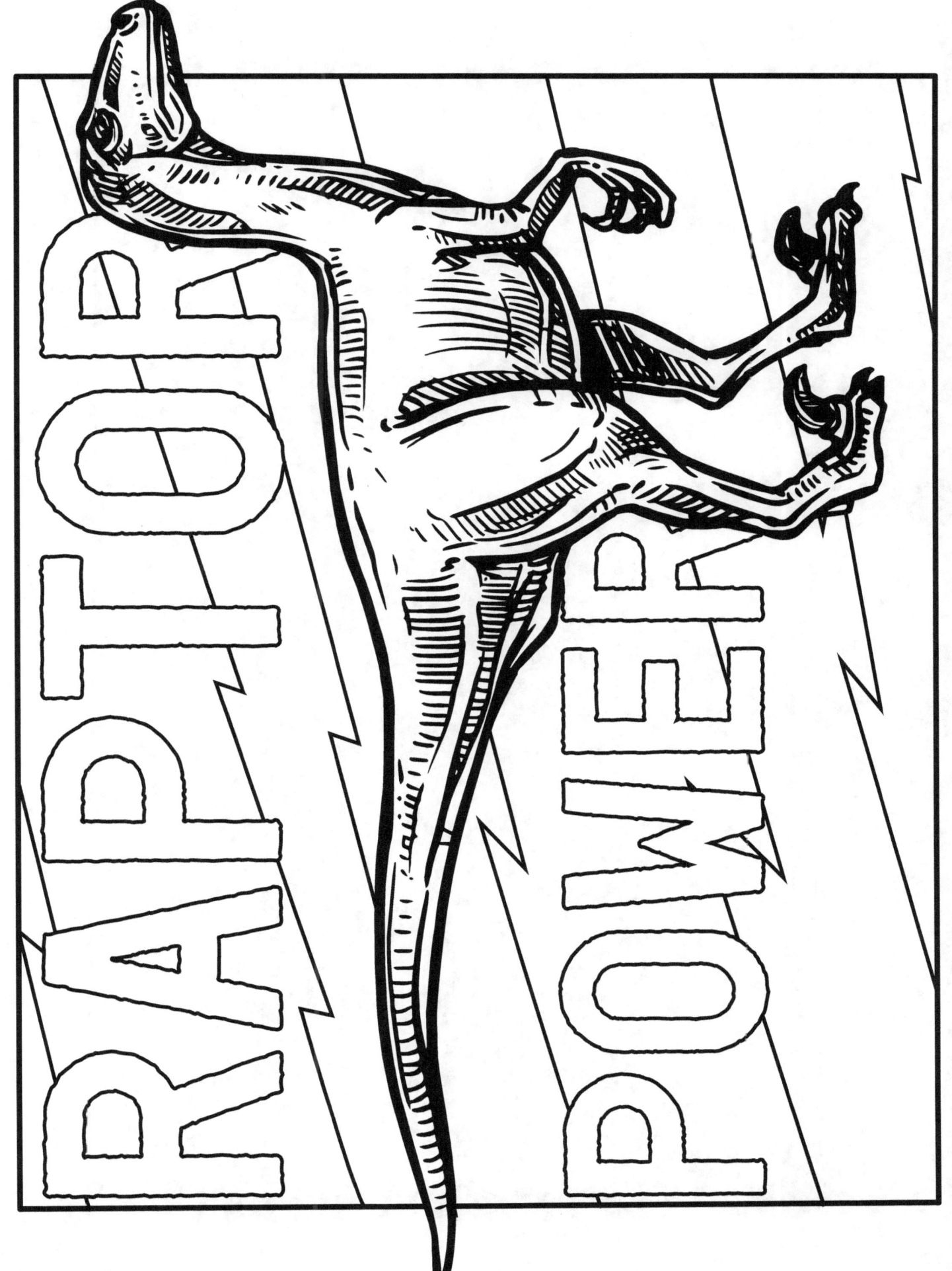

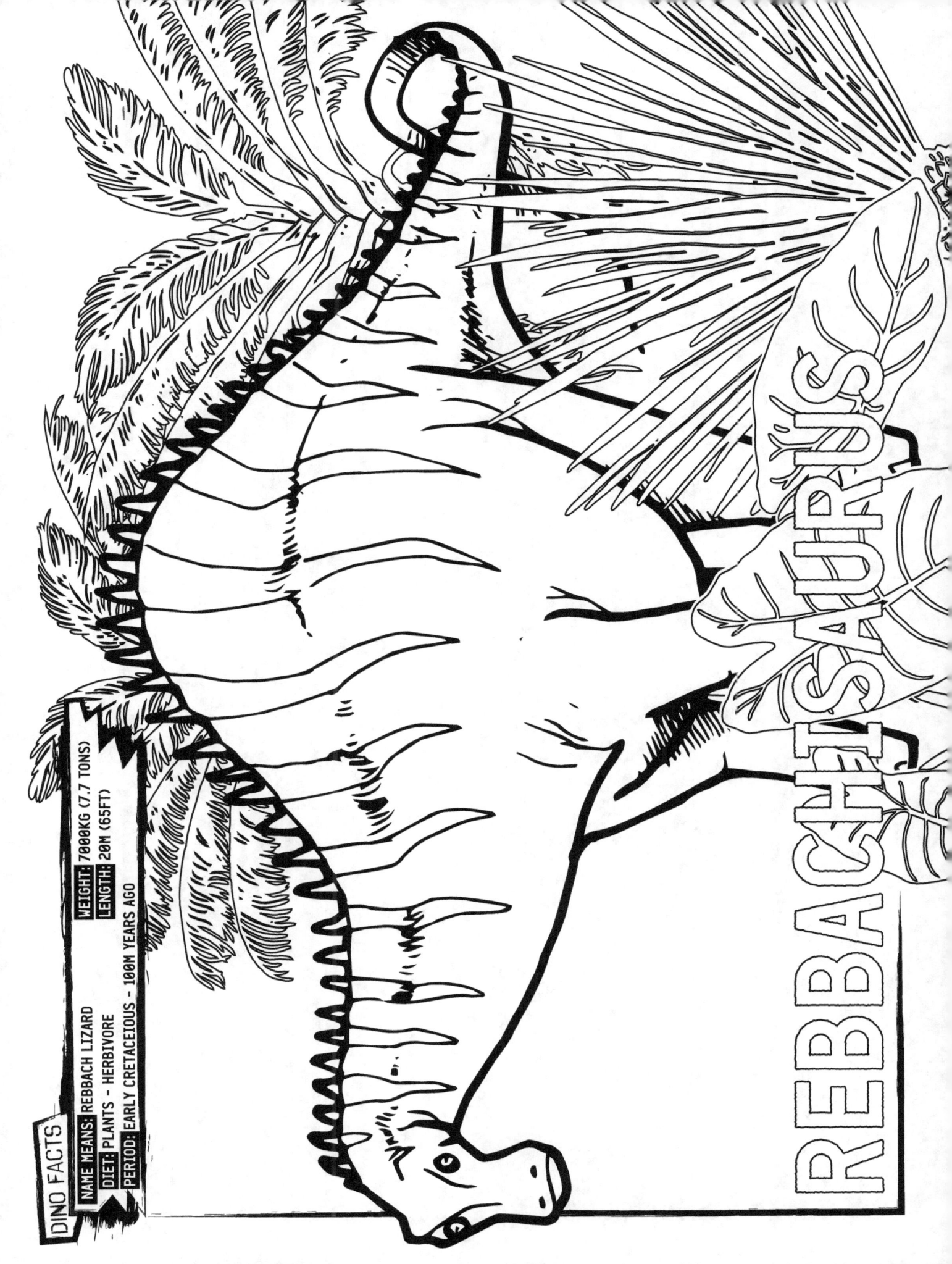

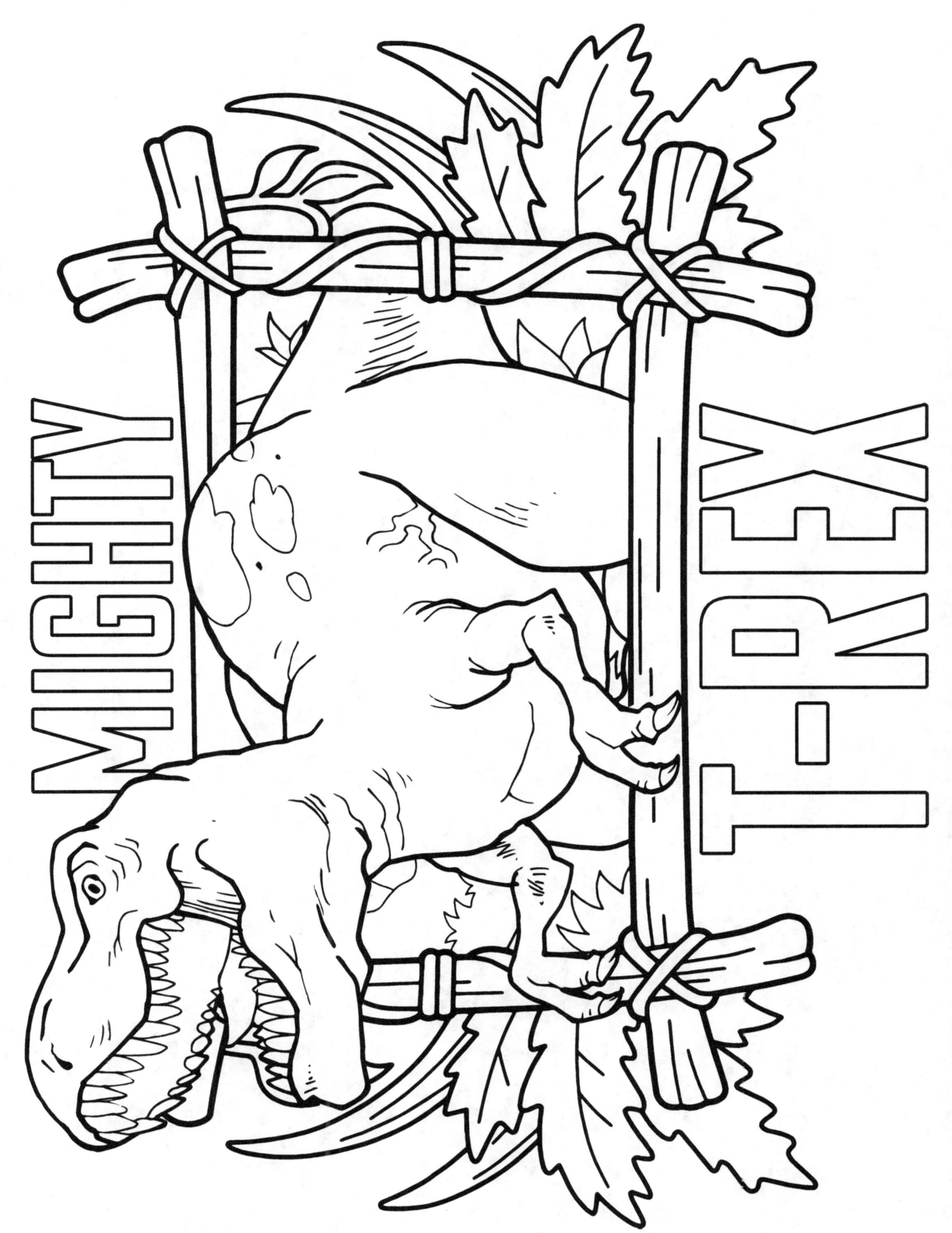

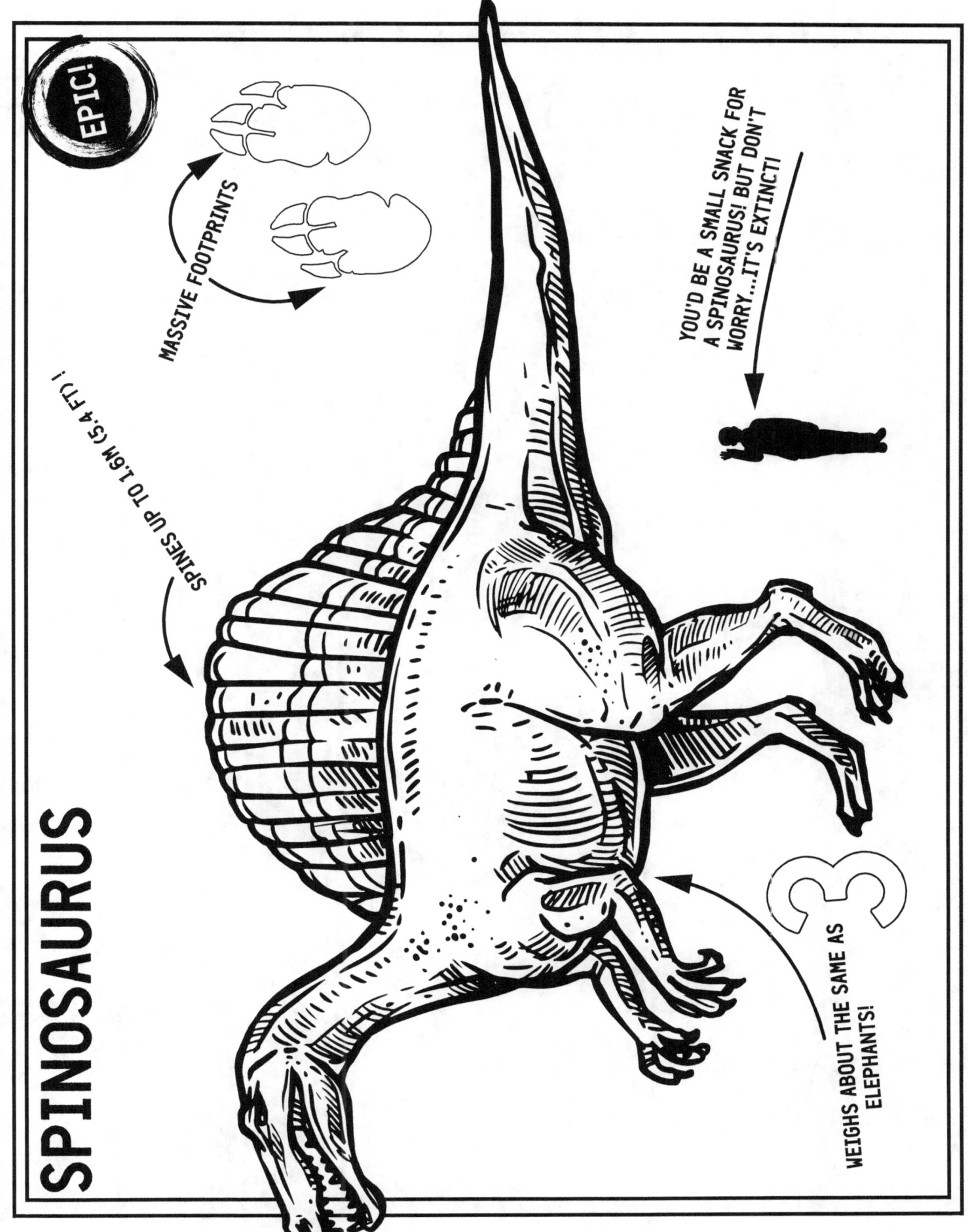

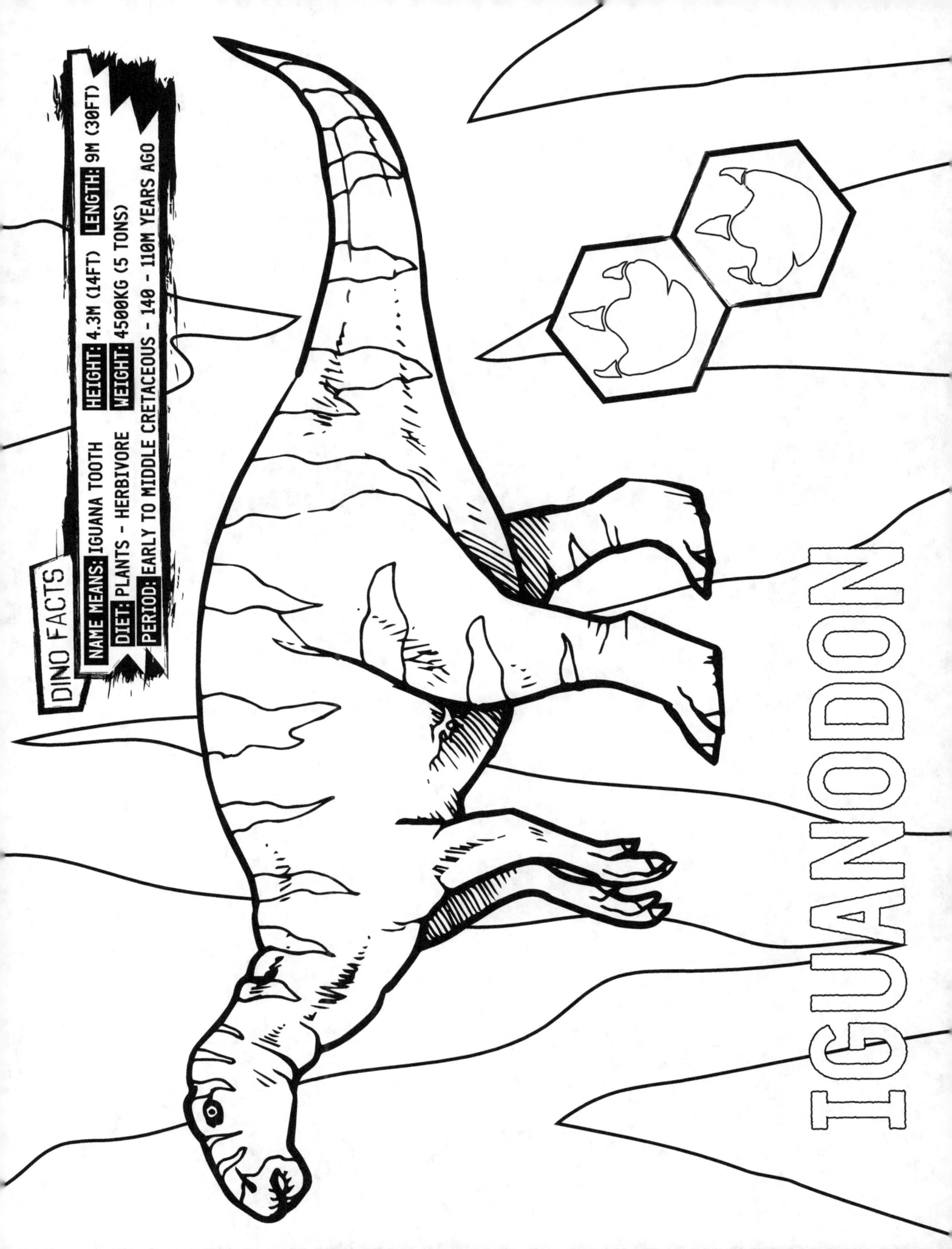

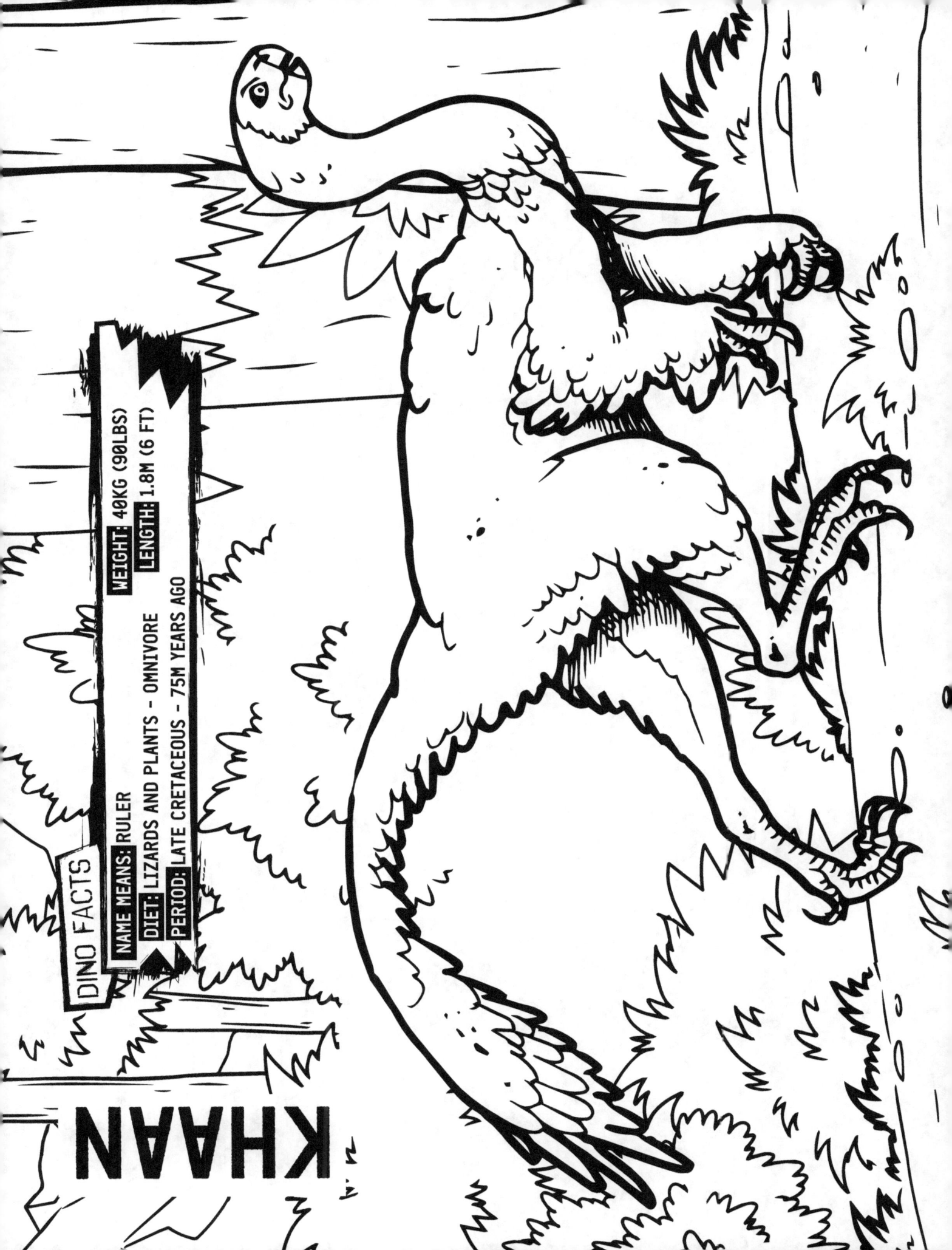

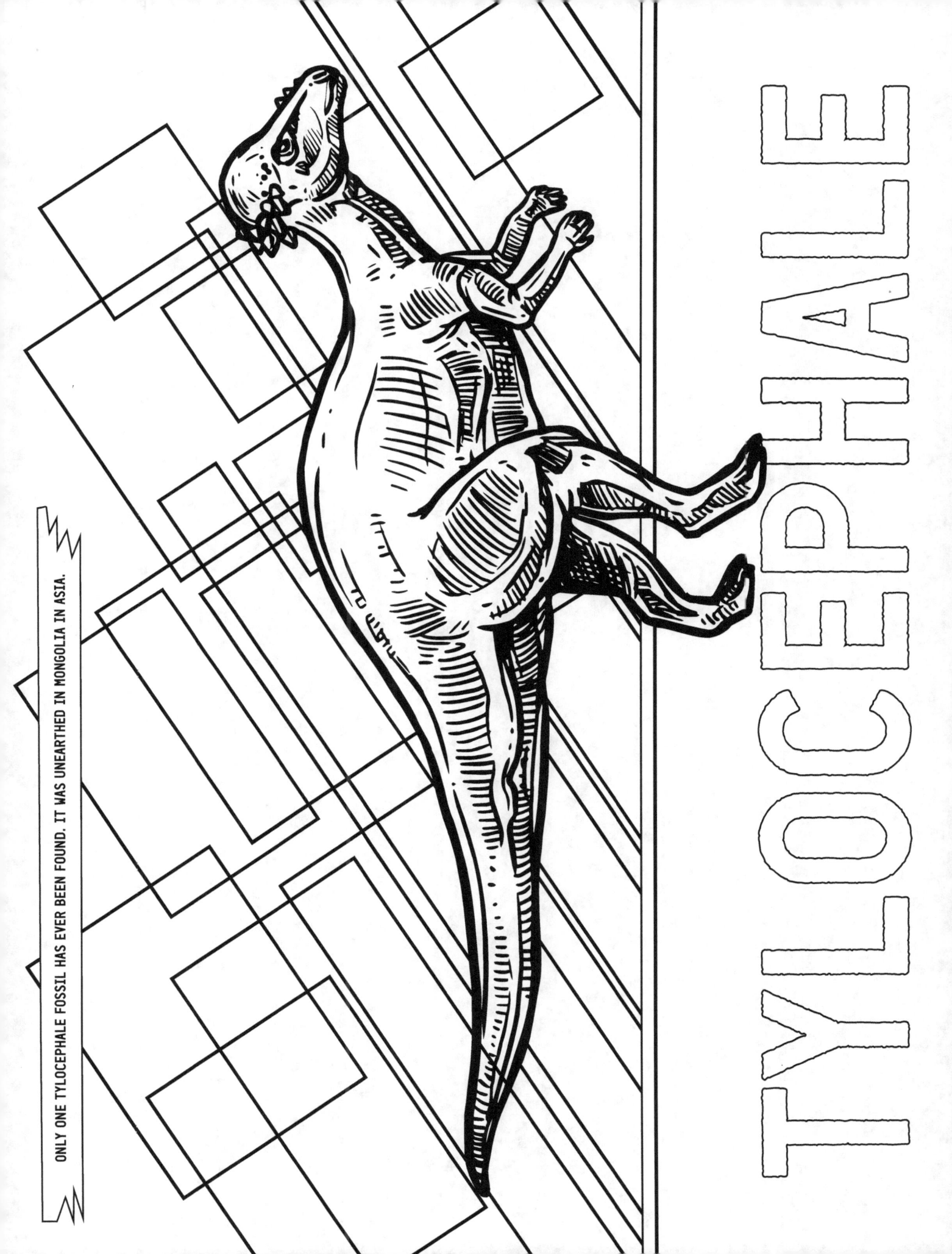

ONLY ONE TYLOCEPHALE FOSSIL HAS EVER BEEN FOUND. IT WAS UNEARTHED IN MONGOLIA IN ASIA.

# TYLOCEPHALE

ANKYLOSAURUS LIVED IN THE WESTERN USA AND CANADA. IT WAS SLOW MOVING AND HEAVY WITH STRONG ARMOR ALL OVER ITS BODY TO PROTECT IT.

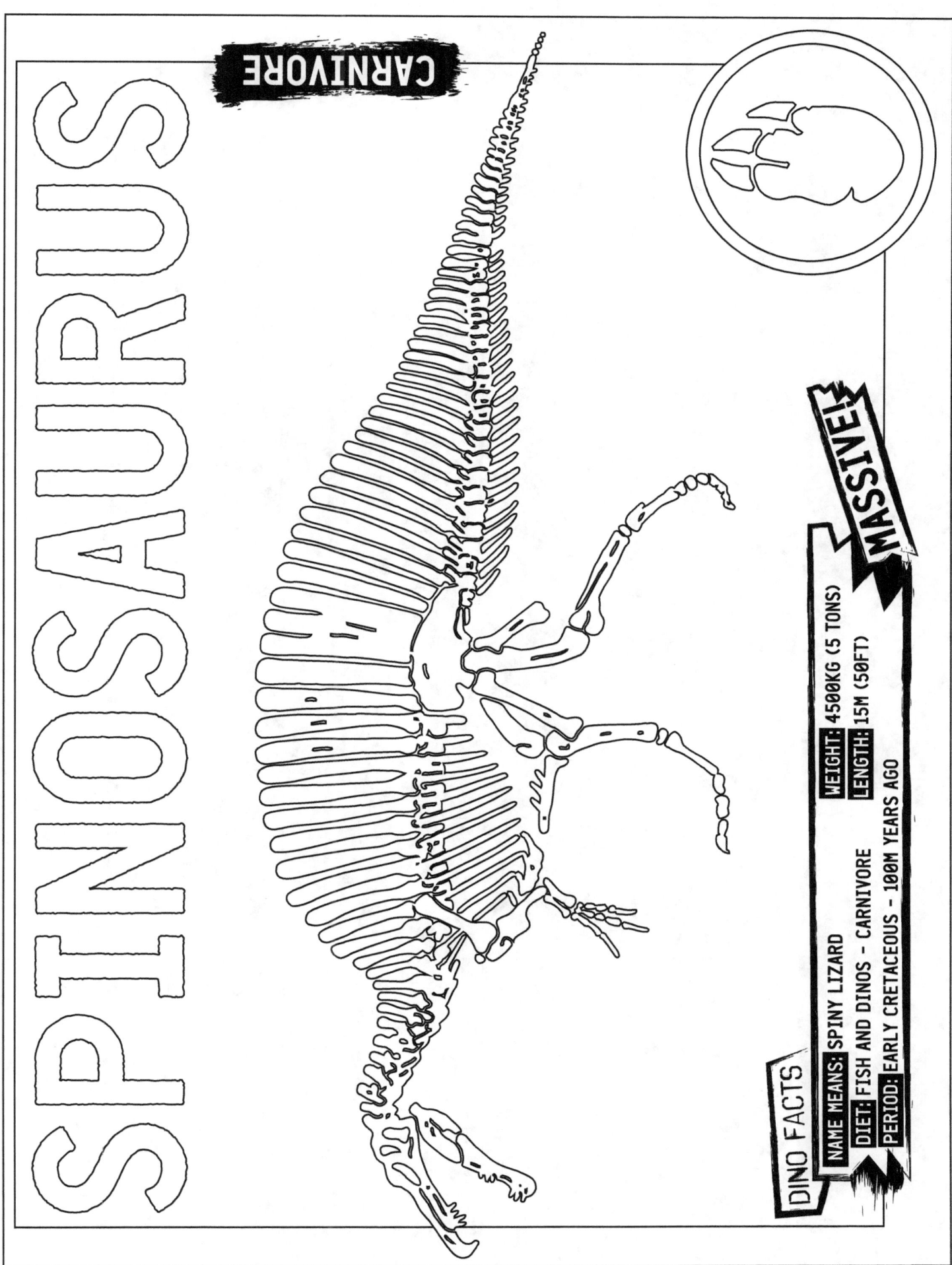

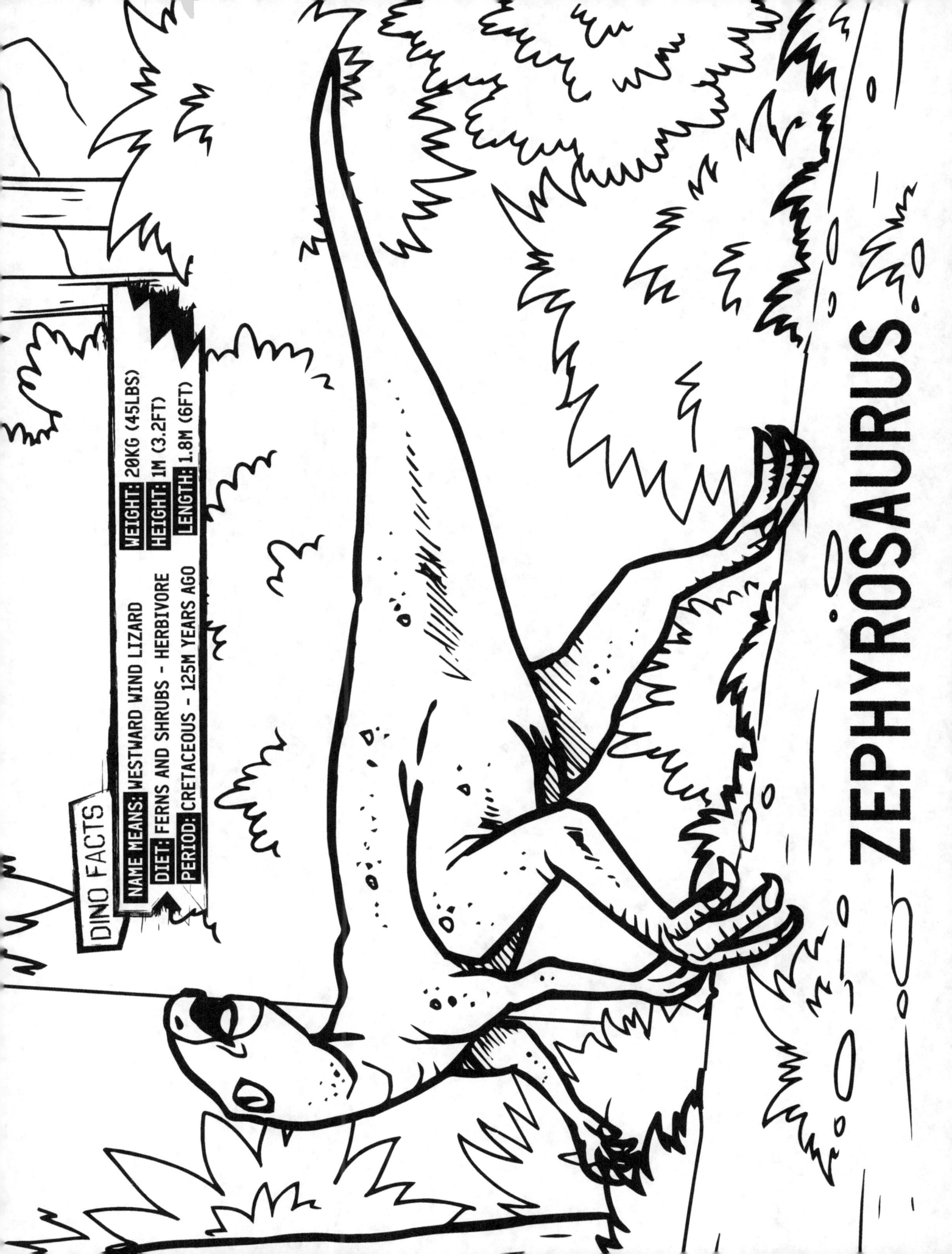

TYRANNOSAURUS AND TYLOCEPHALE

# STEGOSAURUS

## DINO FACTS

**WEIGHT:** 4500KG (5 TONS)
**HEIGHT:** 4.3M (14FT)
**LENGTH:** 9M (30FT)

**NAME MEANS:** ROOF LIZARD
**DIET:** PLANTS - HERBIVORE
**PERIOD:** LATE JURASSIC - 150M YEARS AGO

STEGOSAURUS LIVED IN THE JURASSIC PERIOD AND IT BECAME EXTINCT MORE THAN 60 MILLION YEARS BEFORE TYRANNOSAURUS EVEN EXISTED!

**HERBIVORE**

PARASAUROLOPHUS

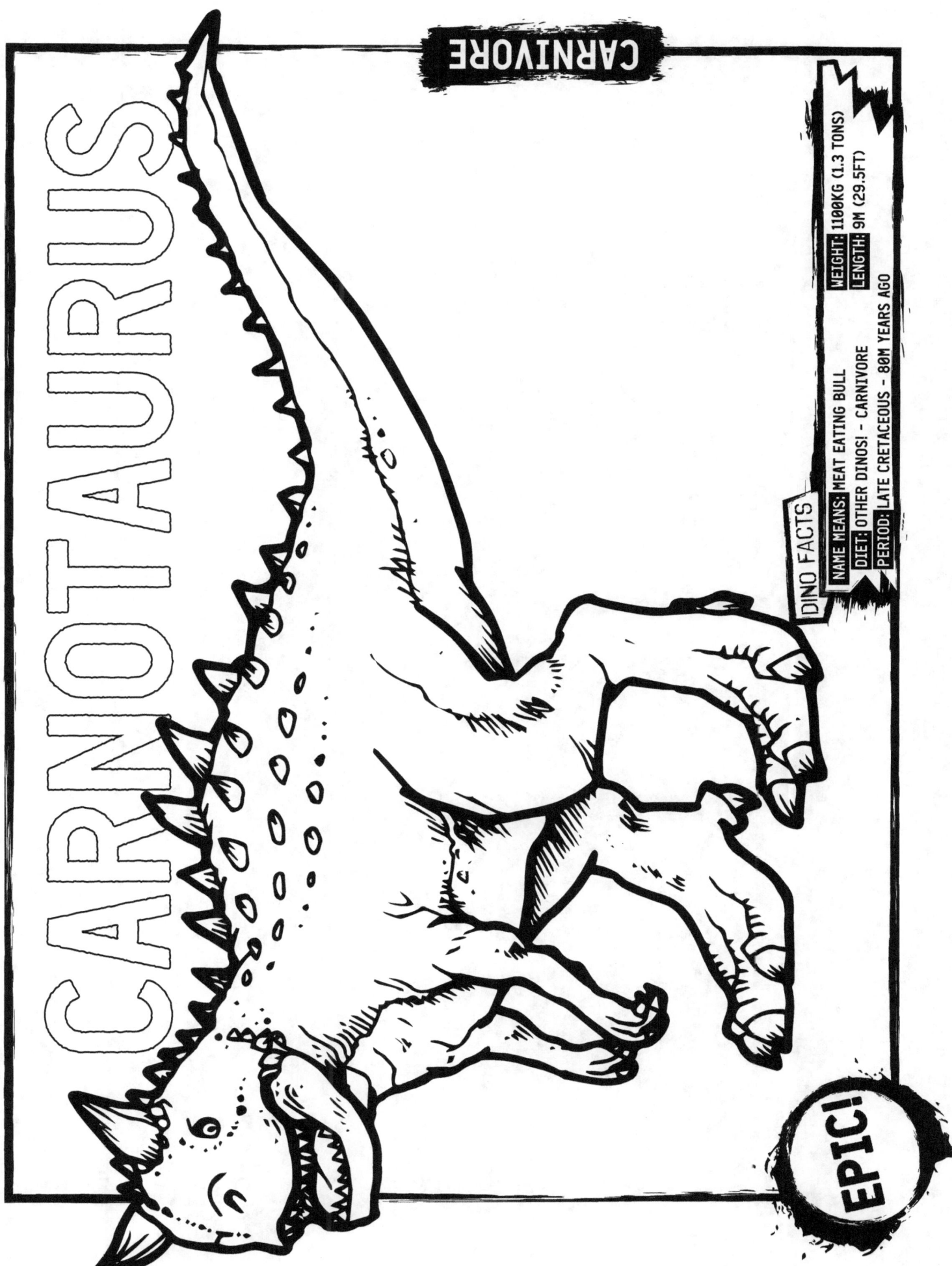

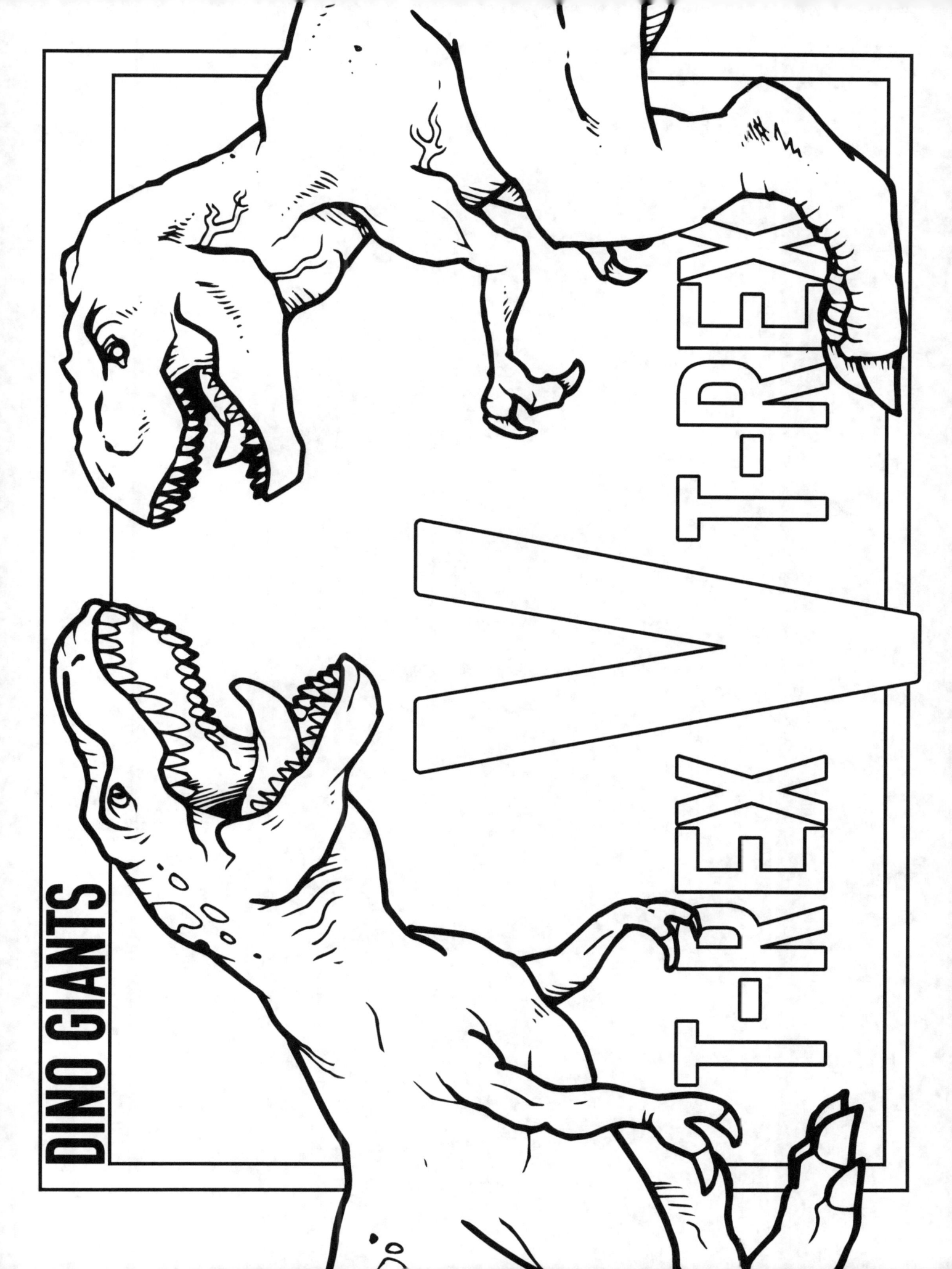

We hope you enjoyed this book. As we learn and grow, we'd love a rating or review for it on Amazon, if you have time. **Thank You!**

Loads more from Under The Cover Press available at amazon

ISBN 979-8552067565

ISBN 979-8509492808

ISBN 979-8520557715

ISBN 979-8590346219

ISBN 979-8695161878

ISBN 979-8575406419

ISBN 979-8559845876

ISBN 979-8559850436

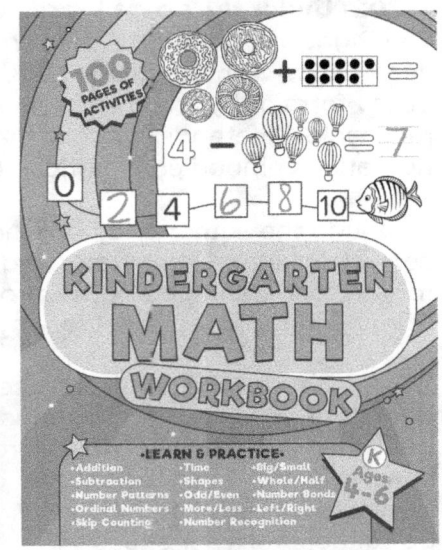

ISBN 979-8717778565

© 2021 Under the Cover Press. All Rights reserved. WWW.UNDERTHECOVERPRESS.COM

The Right of Under the Cover Press to be identified as the author of this work has been asserted by them in accordance with the CDPA 1988. No part of this publication may be reproduced, distributed, or transmitted in any form or by any means, including photocopying, recording, digital scanning, or other electronic or mechanical methods, without the prior written permission of the author, except in the case of brief quotations embodied in critical reviews and certain non-commercial uses permitted by copyright law.

Although the author and publisher have made every effort to ensure that the information in this book was correct at launch, the author and publisher do not assume and hereby disclaim any liability to any party for any loss, damage or disruption caused by errors and omissions, whether such errors or omissions result from negligence, accident, or any other cause.

This content of this book is presented solely for informational purposes. The author and publisher are not offering it as legal, medical, educational, or professional services advice. Neither the author nor the publisher shall be held liable or responsible to any person or entity with respect to any loss or incidental or consequential damages caused, or alleged to have been caused, directly or indirectly, by the information or advice contained herein. As every situation is different, the advice and methods contained herein may not be suitable for your situation.

www.ingramcontent.com/pod-product-compliance
Lightning Source LLC
Chambersburg PA
CBHW080555220526
45466CB00010B/3153